SAMSUNG

C000055763

A12 User Guide

A Complete Manual for Beginners and Seniors with Tips and Tricks to Master the New Galaxy A12 like a Pro

GEORGE THOMAS

Table of Contents

Chapter 1: Basics

Maintaining water as well as dust resistance

Basically, your device can support water as well as dust resistance. Please make sure you follow these tips carefully so as to maintain the water as well as dust resistance of your device. Meanwhile, failure to do so may result to damage to your device.

- If you immerse the device in any liquid other than fresh water, such as salt water, ionised water, or alcoholic beverage, please be sure that the liquid will enter the device faster.
- Please do not expose the device to **water moving with force**.

- Again, if the device is exposed to fresh water, please make sure that you dry it thoroughly with a clean, soft cloth. Also, if the device is exposed to other liquids, such as salt water, swimming pool water, soapy water, oil, perfume, sunscreen, hand cleaner, or chemical products such as cosmetics, please make sure you rinse it with fresh water and dry it thoroughly with a clean, soft cloth. Please note that failure to follow these instructions will lead to the poor performance of the device.

- Please after cleaning the device with a dry cloth, please make sure you dry it thoroughly before making use of it again.

- Please note that the touchscreen as well as other features may not work properly **if the device is been used in water**.

- **In case you don't know, if the device is dropped or receives an impact**, the water- as well as the dust-resistant features of the device may be damaged. We recommend that you handle your device with care.

- Please note that your device has been tested in a controlled environment and as such, it is therefore certified to be water- as well as dust-resistant in specific situations.

Charging the battery

Meanwhile, charge the battery before using it for the first time or when it has been unused for extended periods.

Please make sure you make use of only the Samsung-approved battery, charger, as well as cable that is particularly designed for your device. Please note that incompatible battery, charger, as well as cable can cause serious injuries or damage to your device.

- Do not connect the charger improperly to your device as it may cause serious damage. It should be noted that any damage caused by misuse is not covered by the warranty.
- Please make use of only the USB Type-C cable that is supplied with the device. Please note that the device may be damaged if you use Micro USB cable.

Again, to save energy, please make sure you unplug the charger when it is not in use. Also, the charger does not have a power switch and as such, there is the need for you to unplug the charger from the electric socket when not in use so as to avoid wasting power. Please note that the charger should remain close to the electric socket so that it can be easily accessible while charging.

Wired charging

Please make sure you connect the USB cable to the USB power adaptor as well as plugging the cable into the device's multipurpose jack to charge the battery. Please after it has been fully charge; make sure you disconnect the charger from the device.

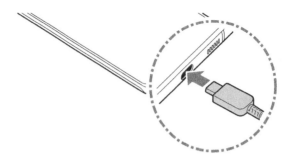

The Wireless charging

Basically, the device has a built-in wireless charging coil. Please note that you can charge the battery through a wireless charger (sold separately).

Please make sure you place the centre of the device's back on the centre of the wireless charger to charge the battery. When it is fully charged, please make sure you disconnect the device from the wireless charger.

Also, the estimated charging time will display on the notification panel. Please note that the actual charging time may vary depending on the charging conditions.

Precautions for wireless charging

- Please don't place the device on the wireless charger when a conductive material, such as metal objects as well as magnets that are placed between the device and the wireless charger. Please note that the device may not charge properly or may overheat, or the device as well as the cards may even be damaged.

- Please do not make use of the wireless charger in areas with weak network signals, based on the fact that you may lose network reception.

- Please always make use of the Samsung-approved wireless chargers. Based on the fact that the battery may not charge properly when you make use other wireless chargers.

Quick charging

Please don't forget to launch the **Settings** app, please make sure you click on the **Device care** → **Battery** → **Charging**, and after that activate the feature you want.

Fast charging: To use the fast charging feature, please make sure you make use of a battery charger that supports Adaptive fast charging.

Fast wireless charging: To make use of this feature, please make sure you use a charger as well as components that support the fast wireless charging feature.

- Please note that you can actually charge the battery more quickly while the device or its screen is turned off.

- Please note that the fan inside the charger may produce noise during fast wireless charging. Also, it is very possible for you to set the fast wireless charging feature to turn off automatically at the preset time when you make use of the **Turn off as scheduled** option. Also as soon as the fast wireless charging feature turns off, the charger's fan noise as well as the indicator light will be reduced.

The Wireless PowerShare

Please note that you can actually charge another device with your device's battery. It is also possible for you to still charge another device even while charging your device. However, depending on the type of accessories or cover being used, the Wireless PowerShare feature may not work properly. Also, it is recommended that you remove any accessories and cover being used before using this feature.

1. Please make sure that you open the notification panel and don't forget to swipe downwards, and then please click on (**Wireless PowerShare**) to activate it.

2. Pease don't forget to place the other device on the centre of your device, with their backs facing.

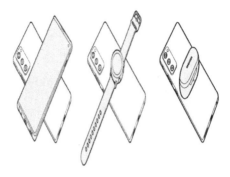

3. Also, when you are finished charging, please make sure you disconnect the other device from your device.

Please do not make use of the earphones while sharing power, based on the fact that you may affect nearby devices by so doing.

- Again, the location of the wireless charging coil may vary by device model. Please don't forget to adjust the devices to connect with each other properly.

- Meanwhile, some features may not available while sharing power.

- Please note that only devices that support the wireless charging feature can be charged when using this feature based on the fact that some devices may not be charged.

- Also, for proper charging, make sure you don't move or use either device while charging.

- Please note that the power charged to the other device may be less compared to the amount shared by your device.

- Also charging other device when charging your device may decrease the charging speed or the device may not even charge properly depending on the charger type.

- Please note that the charging speed or efficiency may vary depending on the device's condition or the surrounding environment.

Setting the limit for power sharing

Meanwhile, please note that you can actually set the device to stop power sharing when the remaining battery power drops below a certain level.

To launch the **Settings** app, please click on **Device care** → **Battery** → **Wireless PowerShare** → **Battery limit**, and after that set the limit.

The Reducing the battery consumption

Please note that your device provides various options that help you conserve battery power.

- Please make sure you optimise the device using the device care feature.
- Also, when you are not using the device, please make sure you turn off the screen by pressing the Side key.
- Please don't forget to activate power saving mode.
- Please make sure you close unnecessary apps.
- Please don't forget to deactivate the Bluetooth feature when not in use.
- Please make sure you deactivate auto-syncing of apps that need to be synced.
- Please don't forget to decrease the backlight time.

- Please do well to decrease the screen brightness.

Battery charging tips and precautions

- Please note that if the battery is completely discharged, the device cannot be turned on instantly when the charger is connected. Please make sure you allow a depleted battery to charge for a few minutes before turning on the device.
- Please when making use of a multiple apps at once, network apps, or even a apps that need a connection to another device, be rest assure that the battery will definitely drain quickly. And as such, to avoid losing power during a data transfer, please make sure that you always make use of these

apps after fully charging the battery.

- Please note that making use of a power source other than the charger, such as a computer, May in no doubt result in a slower charging speed due to a lower electric current.

- Although the device can be used while it is charging, but it may take longer time for the battery to be fully charged.

- Also, if the device receives an unstable power supply while charging, please note that the touchscreen may not function. And as such, when this happens, don't forget to quickly unplug the charger from the device.

- Also, while charging, the device as well as the charger may heat up.

Please note that that is normal and should not affect the device's lifespan or performance. Again, when the battery gets hotter than normal, the charger might stop charging. But if that occurs during wireless charging, please make sure you disconnect the device from the charger in other to allow it to cool down, after that charge the device again later.

- Also, note that when you charge the device while the multipurpose jack is wet, the device may be damaged.

- Please make sure that you dry the multipurpose jack thoroughly before charging the device.

- Again, when the device is not charging properly, please make sure you take the device and the charger to a Samsung Service Centre.

The turning the device on and off

Please do not forget to follow all posted warnings as well as directions from authorised personnel in areas where the use of wireless devices is restricted, such as aeroplanes and hospitals.

Please don't forget to press as well as hold the Side key for a few seconds to turn on the device.

The turning the device off

1. Please do not forget to turn off the device, press and hold the Side key as well as the Volume Down key simultaneously. Alternatively, open the notification panel and tap ⏻.

2. Please make sure you click on the **Power off**.

Also to restart the device, please click on **Restart**.

Please note that you can set the device to turn off when you press as well as hold the Side key. Please make sure you launch the **Settings** app, don't forget to click on the **advanced features** → **Side key**, and after that please click on the **Power off menu** under **Press and hold**.

Forcing restart

Again, when your device is frozen and unresponsive, please make sure you press and hold the Side key and the Volume Down key simultaneously for more than 7 seconds to restart it.

Emergency mode

Meanwhile, you can also switch the device to emergency mode to you can even reduce battery consumption although, some apps as well as functions will be restricted. Again, in the emergency mode, you can actually make an emergency call, send your current location information to others as well as , sounding an emergency alarm among others.Meanwhile, to activate emergency mode, please don't forget to press and hold the Side key and the Volume Down key simultaneously, and after that please make sure you click on the **Emergency mode**. Also, don't forget to also open the notification panel as well as clicking on the **emergency mode**.

To deactivate emergency mode, please don't forget to click on the ⋮ → **Turn off Emergency mode**.

Meanwhile, the usage time left shows the time remaining before the battery power runs out. Again, please note that the usage time left may vary depending on your device settings and operating conditions.

Initial setup

Again, as soon as you must have turn on your device for the first time or after performing a data reset, please make sure you follow the on-screen instructions to set up your device.

Failure to connect to a Wi-Fi network, may serve as a hindrance for you to set up some device features during the initial setup.

Samsung account

Please note that your Samsung account is an integrated account service that permits you to make use of a variety of Samsung services that is provided by mobile devices, TVs, as well as the Samsung website.

1. Please don't forget to launch the **Settings** app as well as clicking on the **Accounts and backup** → **Accounts** → **Add account** → **Samsung account**.

2. Please make sure you alternatively, launch the **Settings** app along with clicking on it 🔵.

3. Again, if you already have a Samsung account, please make sure you sign in to your Samsung account.

   • Also, if you want to sign in using your Google account, please make sure you click on **Continue with Google**.

- But if you don't have a Samsung account, please click on **Create account**.

Finding your ID and resetting your password

In case you forget your Samsung account ID or password, please click on **Find ID** or **Reset password** on the Samsung account sign-in screen. Also, it is very possible for you to locate your ID or reset your password after you must have entered the required information.

Signing out of your Samsung account

Please, it should be noted that when you sign out of your Samsung account, your data, such as contacts or events, will also be removed from your device.

1. Please make sure you launch the **Settings** app as well as clicking on the **Accounts and backup** → **Accounts**.

2. Please don't forget to click on the **Samsung account** → **Personal info** → **⋮** → **Sign out**.

3. Please make sure you click on **Sign out**, enter your Samsung account password, and after, please click on OK.

Transferring data from your previous device (Smart Switch)

Meanwhile, you can also make use of the Smart Switch to transfer data from your previous device to your new device.

To launch the **Settings** app please click on **Accounts and backup** → **Smart Switch**.

- Although this feature might not be supported on some devices or computers but Samsung only transfer content that you own or have the right to transfer.

Transferring data wirelessly

The following are some of the ways you can transfer data from your previous device to your device wirelessly via Wi-Fi Direct:

1. Meanwhile, on the earlier device, please make sure you launch **Smart Switch**.

2. But if you don't have the app, please make sure you download it from **Galaxy Store** or **Play Store**.

3. Again, on your device, please make sure you launch the **Settings** app and please don't forget to click on

the **Accounts and backup** → **Smart Switch**.

4. Please make sure you place the devices near each other.

5. Also, on the previous device, please make sure you click on **Send data** → **Wireless**.

6. Again, in the previous device, please make sure you click on **Allow**.

7. Also, in your device, please don't forget to select an item to bring and also make sure you click on Transfer.

Backing up and restoring data using external storage

Also when you want to transfer data using external storage, such as a microSD card, please make sure you take the following into consideration;

1. Please make sure you back up data from your previous device to external storage.

2. Please don't forget to insert or connect the external storage device to your device.

3. Also, on your device, please make sure you launch the **Settings** app as well as clicking on the **Accounts and backup** → **Smart Switch** → 🗔 → **Restore**.

4. Please make sure you follow the on-screen instructions to transfer data from external storage.

Transferring backup data from a computer

When transferring data between your device and a computer, please note that there is the need for you to first of all download the Smart Switch computer version app from www.samsung.com/smartswitch before backing it up from your previous device to a computer and import the data to your device.

1. Also, in the computer, please make sure you visit www.samsung.com/smartswitch to download Smart Switch.

2. In the computer, please don't forget to launch Smart Switch.

   • In case the previous device is not a Samsung device, please make sure you back up the data to a computer using a program that is provided by the device's manufacturer. After that, please make sure you skip to the fifth step.

3. Please don't forget to connect your previous device to the computer using the device's USB cable.

4. Also, in the computer, please make sure you follow the on-screen instructions so as to back up data from the device.

5. After that, please make sure you disconnect your previous device from the computer.

6. Please don't forget to also connect your device to the computer using the USB cable.

7. Meanwhile, in the computer, please make sure you follow the on-screen instructions to transfer data to your device.

Changing the screen lock method

Again, in respect to changing the screen lock method, please make sure you launch the **Settings** app, click on **Lock screen** → **Screen lock type**, and after that select a method.

Please note that it is very possible for you to protect your personal information by preventing others from accessing your device when you set a pattern, PIN, password, or your biometric data for the screen lock method. Please note that after setting up the screen lock method, the device will always require an unlock code whenever you want to acess it.

Meanwhile, it is very possible for you to set your device to perform a factory data reset should in case you enter the unlock code incorrectly for several times in a row in addition to you reaching the attempt limit. Please make sure you launch the **Settings** app, please don't forget to also click on the **Lock screen** → **Secure lock**

settings, unlock the screen using the preset screen lock method, and after that, please make sure you click on the **Auto factory reset** switch to activate it.

Indicator icons

More so, the indicator icons will appear on the status bar at the top of the screen. The following are the most common icons listed in the table below.

- Also, the status bar may not appear at the top of the screen in some apps. And as such, to have it display the status bar, please make sure you drag down from the top of the screen.

- Please note that some indicator icons appear only when you open the notification panel.

- Also, the indicator icons may appear in a different way

depending on the service provider or model.

Controlling media playback

Please note that you can easily access music or video playback when using the Media feature. And it is also very possible for you to continue playback on another device.

1. Please open the notification panel and don't forget to click on **Media**.

2. Please make sure you click on the icons on the controller to control the playback.

3. Meanwhile, to continue playback on another device, please click on ⊚ and choose the device you want.

Controlling nearby devices

Please make sure you launch it as quickly as you can so that you can take control of the nearby connected devices as well as frequently used SmartThings devices and scenes on the notification panel.

The following are the ways you can control media playback:

1. Please make sure you open the notification panel and don't forget to click on **Devices**.

2. Please note that nearby connected devices as well as SmartThings devices and scenes will appear.

3. Please don't forget to choose a nearby device or a SmartThings device to control it, or you can as well choose a scene to launch it.

In case you receive a notification with edge lighting, please note that you can quickly view its content as well as performing available actions by opening the pop-up window. For instance, if you receive a message at the same time that you are watching a video or playing a game, please note that it is very possible for you to view the message as well as reply to it without switching the screen.

Also, when you receive a notification with edge lighting when making use of the app, please make sure you drag the notification downwards.

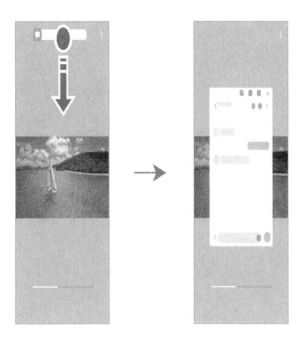

Please note that this feature is only available to apps that support the Multi window as well as edge lighting features. Meanwhile, to view supported apps, please make sure you launch the **Settings** app, thereafter click on **Display** → **Edge screen** → **Edge lighting** → **Choose apps**.

Screen capture and screen record

Screen capture

Furthermore, you can capture a screenshot while using the device in addition to write on, draw on, crop, or even share the captured screen as the case maybe. It's also possible for you to capture the current screen and scrollable area.

How to capture a screenshot

The following methods can be use to capture a screenshot that you can view in **Gallery**:

Method 1) Key capture: Please make sure you press the Side key and the Volume Down key simultaneously.

Method 2) Swipe capture: Please don't forget to swipe your hand to the left or right across the screen.

- Please note that it is not possible to capture a screenshot when using some apps and features.

- Also, if capturing a screenshot by swiping is not activated, please make sure you launch the **Settings** app, please click on **Advanced features** → **Motions and gestures**, and after that please make sure you click on the **Palm swipe to capture** switch to activate it.

Once you have captured a screenshot, please make sure you make use of the following options on the toolbar at the bottom of the screen:

⌞⌄⌝ : Please don't forget to capture the current content as well as the hidden content on an elongated page, such as a webpage. Also, when you tap⌜⌄⌟, note that the screen will automatically scroll down and as such, more content will be captured.

⏿ : Please make sure you write or draw on the screenshot or crop a portion from the screenshot. It is also very possible for you to view the cropped area in **Gallery**.

: Please don't forget to add tags to the screenshot. Again, in case you want to search for screenshots through tag, please click on **Search** located at the top of the Apps screen. That way, you can view the tags list and easily search for the screenshot that you want.

⌾ : Please make sure you share the screenshot with others.

In case the options are not visible on the captured screen, please make sure you launch the **Settings** app, please click on **Advanced features → Screenshots and screen recorder**, and after that please click on the **Screenshot toolbar** switch to activate it.

Screen record

The following are the ways you can record the screen while using your device:

1. Please don't forget to open the notification panel, swipe downwards, and after that please click on (**Screen recorder**) to activate it.

2. Please don't forget to click on the sound setting as well as clicking on the **Start recording**. Please note that after the countdown, recording will start.

- Also if you want to write or draw on the screen, please click on ✏.

- In case you want to record the screen with a video overlay of yourself, please click on ♟.

3. Also, when you are done recording the video, please click on ■.

You can view the video in **Gallery**.

In case you want to change the screen recorder settings, please make sure you launch the **Settings** app and after that lease click on **advanced features** → **Screenshots and screen recorder** → **Screen recorder settings**.

Chapter 2: Apps and features

Installing or uninstalling apps

Galaxy Store

In terms of purchase and download apps , please note that you can actually download the apps that are specialised for Samsung Galaxy devices.

To launch the **Galaxy Store** app, please don't forget to browse apps by category or you can simply click Q to search for a keyword.

• To change the auto update settings, click on → → **Auto update apps**, and then select an option.

Play Store

The following are ways you can purchase and download apps.

Meanwhile, you can launch the **Play Store** app simply by browsing through the apps by category or search for apps by keyword.

Also, to change the auto update settings, please make sure you click on → **Settings** → **Auto-update apps**, and after that choose an option.

Managing apps

Uninstalling or disabling apps

Please don't forget to click and hold an app in addition to selecting an option.

- **Uninstall**: Please note that it is very possible for you to uninstall downloaded apps.
- **Disable**: Also, it is very possible for you to disable selected default apps that cannot be uninstalled from the device.

Basically, some apps might not support this feature.

Enabling apps

Please don't forget to launch the **Settings** app, and click on **Apps** → ▼ → **Disabled**, please make sure you choose the app, and after that click on **Enable**.

Setting app permissions

Also, for some apps to operate properly, they might not be needing permission to access or make use of the information on your device.

Also, for you to actually view your app permission settings please launch the **Settings** app as well as clicking on the **Apps**. Select an app. Also, please don't forget to click on **Permissions**. Meanwhile, it is very possible for you to view the app's permissions list in addition to changing its permissions.

Again, for you to view or change app permission settings by permission category, please launch the **Settings** app and click on the **Apps** → ⋮ → **Permission manager**. Please make sure you choose an item as well as select an app.

In case you don't have permissions to apps, please be inform that the basic features of the apps may not function properly.

Phone

Introduction

To make or answer voice as well as video calls.

Please note that if the area around the rear camera is covered, unwanted noises may incur during a call and as such, there is the need for you to do away with accessories, such as the screen protector or stickers that are around the rear camera .

Making calls

The following are the ways you can make calls:

1. Please launch the **Phone** app and click on the **Keypad**.

2. Please enter a phone number.

3. Please don't forget to click on 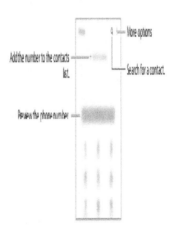 to make a voice call

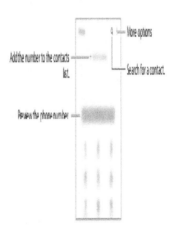

Making calls from call logs or contacts list

In case you want to make calls from call log or contact list, please make sure you launch the **Phone** app, click on the **Recents** or **Contacts**, and after that don't forget to swipe to the right on a contact or a phone number so as to make a call.

In case this feature is deactivated, please make sure you launch the **Settings** app, click on **Advanced features** → **Motions and gestures**, and after that click to **Swipe to call or send messages** switch to activate it.

Using speed dial

Furthermore, to set a number to speed dial, please make sure you launch the **Phone** app, click on **Keypad** or **Contacts** → **⋮** → **Speed dial numbers**, please don't forget to also choose a speed dial number, and after that add a phone number.

Meanwhile, for you to make a call there is the need for you to click and hold a speed dial number on the keypad. Also, for speed dial numbers 10 and up, please don't forget to click the first digit(s) of the number, and after that click and hold the last digit.

For instance, if you set the number 123 as a speed dial number, please make sure you click on 1, 2, and after that please click and hold 3.

Making calls by searching for nearby places

In case you want to launch the **Phone** app, please click on the **Places**, and after that choose a category or you can as well click on Q and enter a business name in the search field. Otherwise, you can decide yo choose one from the recommended hot places. Please note that the business's information, such as its phone number or address, will appear.

Please note that this feature may not be available depending on the service provider or model.

Making an international calls

The following are the ways you can make international calls:

1. Please launch the **Phone** app and don't forget to click on **Keypad**.

2. Please make sure you click and hold **o** until the + sign appears.

3. Please don't forget to enter the country code, area code, as well as the phone number before clicking on .

Receiving calls

Answering a call

This can be achieve when a call comes in, drag outside the large circle.

Rejecting a call

Again, this can also be achieve when a call comes in, drag outside the large circle.

In case you want to send a message when rejecting an incoming call, please drag the **Send message** bar upwards and choose a message that you desire to send.

Also, in case you want to create various rejection messages, please launch the **Phone** app, and click on ⋮ → **Settings** → **Quick decline messages**, enter a message, and after that please click on ✛.

The following are some of the ways you can actually lock calls from specific numbers added to your block list:

1. Please make sure you launch the **Phone** app after that, please clcik on ⋮ → **Settings** → **Block numbers**.

2. Please don't forget to click on **Recents** or **Contacts**, choose contacts or phone numbers, and after that please click on **Done**. In case you want to manually enter a number, please click on **Add phone number**, don't forget to also enter a phone number, and after that please click on ┼.

Also, when blocked numbers try to contact you, please be informed that you will not receive notifications based on the fact that the calls will be logged in the call log.

Furthermore, it is very possible for you to block incoming calls from people that do not show their caller ID. Please don't forget to click on the **Block unknown/hidden numbers** switch to activate the feature.

Options during calls

The following otions are available during calls:

- **Add call**: Dial a second call. Please not that it is the first call that will be put on hold but when you end the second call, the first call will be resumed.

- **Hold call**: Please make sure you hold a call when the needs arise.

- **Bluetooth**: Please don't forget to always switch to a Bluetooth headset if it is connected to the device.

- **Speaker**: Please make sure you activate or deactivate the speakerphone. Please don't forget to keep the device away from your ears when using the speakerphone.

- **Mute**: Pleas don't forget to turn off the microphone so that the other party cannot hear you.

- **Keypad** / **Hide**: Please make sure you open or close the keypad.

- 📞: End the current call.

- **Camera**: Also, during a video call, you can turn off the camera if you don't want other party to see you.

- **Switch**: Again, during a video call, it is possible for you to switch between the front and rear cameras.

Introduction

Basically, it is possible for you to create new contacts or manage contacts on the device.

Adding contacts

Creating a new contact

You can create new contacts the following ways:

1. Please make sure you launch the **Contacts** app and click on .

2. Please don't forget to choose a storage location.

3. Please make sure you enter contact information and click on **Save**.

Importing contacts

Please note that you can add contacts by importing them from other storages to your device.

1. Pleas launch the **Contacts** app and don't forget to click on $\equiv$ → **Manage contacts** → **Import or export contacts** → **Import**.

2. Please make sure you follow the on-screen instructions to import contacts.

Syncing contacts with your web accounts

The following are some of the ways you can sync your device contacts with online contacts saved in your web accounts, such as your Samsung account.

1. Please launch the **Settings** app and thereafter click on **Accounts and backup** → **Accounts**.Please

doesn't forget to choose the account to sync with.

2. Please make sure you click on **Sync account** and after that, please click on the **Contacts** switch to activate it.

In case you are using the Samsung account, please click on **⋮** → **Sync settings** thereafter click on the **Contacts** switch to activate it.

Searching for contacts

Also, to launch the **Contacts** app. Please make sure you click on Q at the top of the contacts list and enter search criteria.

Please don't forget to also click on the contact. After that take one of the following actions:

☆: Please make sure you add to favourite contacts.

: Please don't forget to make a voice call.

: Please make sure you make a video call.

: Please do well to compose a message.

: Please don't forget to compose an email.

Deleting contacts

The following are the ways you can actually delete contacts:

1. Please launch the **Contacts** app along with clicking on ⋮ → **Delete**.

2. Please make sure you choose a contacts as well as clicking on **Delete**.

In case you want to delete contacts one by one, please click on a contact from the contacts list and click on ⋮ → **Delete**.

Sharing contacts

Again, the following are some of the ways you can actually share contacts with others:

1. Please make sure you launch the **Contacts** app and click on ⋮ → **Share**.
2. Please don't forget to choose contacts and tap **Share**.
3. Please make sure you choose a sharing method.

Creating groups

The following are ways you can add groups, such as family or friends, and manage contacts by group.

1. Please launch the **Contacts** app and after that, please click on ☰ → **Groups** → **Create group**.

2. Please make sure you follow the on-screen instructions to create a group.

Merging duplicate contacts

In case your contacts list includes duplicate contacts, it is very possible for you to merge them into one to streamline your contacts list through:

1. Please launch the **Contacts** app and after that, please click on ☰ → **Manage contacts** → **Merge contacts**.
2. Please make sure you tick contacts and don't forget to **merge**.

Messages

Introduction

Meanwhile, you can actually send and view messages by conversation.

Although you may incur additional charges for sending or receiving messages when you are roaming.

1. Please make sure you launch the **Messages** app and after that, please click on 💬 .

2. Please don't forget to also add recipients and enter a message.

3. In case you want to record and send a voice message, please click and hold ᴵᴵᴵᴵ, say your message, and after that release your finger. Pleasse note that the recording icon only appears while the message input field is empty.

4. Please don't forget to click on 🧭 to send the message.

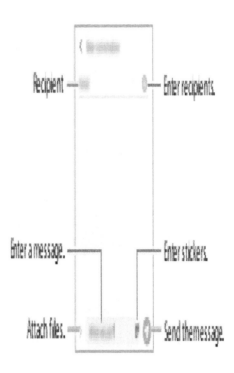

Viewing messages

1. Please make sure you launch the **Messages** app and tap **Conversations**.

2. Again, on the messages list, please choose a contact or a phone number.

- In case you want to reply to the message, please click on the message input field, enter a message, and after that click on .

- In case you want to adjust the font size, please spread two fingers apart or pinch on the screen.

Sorting messages

The following are some of the ways you can sort messages by category and manage them easily.

Please launch the **Messages** app and after that, please click on **Conversations** → **New category** → **Add category**. In case the category of option does not appear, please click on ⋮ → **Settings** and after that, please click on the **Conversations categories** switch to activate it.

Deleting messages

In case you want to delete messages, please click and hold a message to delete, then tap **Delete**.

Changing message settings

To launch the **Messages** app, please click on ⋮ → **Settings**. Please note that it is also possible for you to block unwanted messages, change notification settings, and more.

Internet

Again, you can also browse the Internet to search for information and bookmark on your favourite webpages to access them conveniently:

1. Please make sure you launch the **Internet** app.

2. Please make sure you enter the web address or a keyword, and after that, please click on Go.

In case you want to view the toolbars, please make sure you drag your finger downwards slightly on the screen.

Also, to switch between tabs quickly, please s swipe to the left or right on the address   field

It is very possible for you to set a password so as to prevent others from viewing your search history, browsing history, bookmarks, as well as saved pages.

1. Please make sure you click on 🗇 → **Turn on Secret mode**.

2. Please don't forget to click on the **Lock Secret mode** switch to activate it, also click **Start**, and after that make sure you set a password for secret mode.

Again, in the secret mode, please be rest assured that the device will change the colour of the toolbars. And so, if you want to deactivate secret mode, please click on 🗇 → **Turn off Secret mode**.

Again, in the secret mode, please note that you cannot use some features, such as screen capture.

Introduction

Furthermore, you can take a photos as well as a record videos when using various modes as well as settings.

Camera etiquette

- Please do not take photos or record videos of other people without their permission.
- Please make sure you don't take photos or record videos where legally prohibited.
- Please don't take photos or record videos in places where you may violate other people's privacy.

Taking photos

1. You start by launching the **Camera** app.

Again, you can also launch the app simply by pressing the Side key twice quickly or by dragging it outside the circle on the locked screen.

- Please note that some camera features may not be available when the camera app is from the locked screen or when the screen is turned off when the screen lock method is set.

- Also, the camera can automatically shuts off when unused.

- Again, some methods may not be available depending on the service provider or model.

2. Please don't forget to click on the image on the preview screen where the camera should focus.

Also, to adjust the brightness of photos, please make sure you click on the screen.

When the adjustment bar appears, please make sure you drag the adjustment bar towards.

3. Please don't forget to click ◯ to take a photo.

Also, to change the shooting mode, please drag the shooting modes list to the left or right, or you can as well swipe to the left or right on the preview screen.

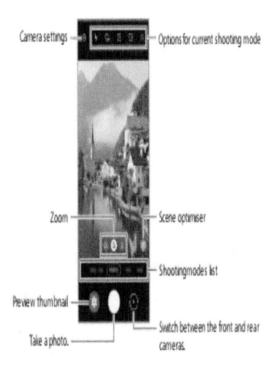

Camera settings — Options for current shooting mode

Zoom — Scene optimiser

Shooting modes list

Preview thumbnail

Take a photo. — Switch between the front and rear cameras.

- Again, the preview screen may vary depending on the shooting mode as well as the camera that is being used.

- In case the photo you take appears blurry, please make sure you clean the camera lens and try again.

- Please ensure that the lens is not damaged or contaminated. If not, the device may not work properly in some modes that require high resolutions.
- Please note that the maximum capacity for recording a video may vary depending on the resolution.

Using zoom features

Please don't forget to choose ✿ ✿ ✿ or drag to the left or right so as to zoom in or out. On the other hand, please make sure you spread two fingers apart on the screen to zoom in, and touch to zoom out. In case the zoom ratio exceeds a certain level, please be rest assured that a zoom guide map will appear to indicate where you are zooming in on the image.

1. Again, the Ultra wide camera permits you to take a wide-angle photos or record wide-angle videos of things like landscapes.

2. Meanwhile, the wide-angle camera also permits you to take basic photos or record normal videos.

3. On the other hand, the telephoto camera permits you to take photos or record videos by enlarging the subject. Zoom features are available only when using the rear camera.

Using the camera button

The following are some of the ways that you can make use of the camera button:

- Please click and hold the camera button to record a video.

- Please don't forget to take burst shots, also swipe the camera button to the edge of the screen and hold it.

- In case you add another camera button, please note that it is very possible for you to move it anywhere on the screen as well as taking photos more conveniently. Again, on the preview screen, please click on the ⚙ → **shooting methods** and don't forget to click on the **Floating Shutter button so as to** switch to activate it.

Photo mode

Also, the camera can adjust the shooting options automatically depending on the surroundings that permit it to capture photos easily.

Meanwhile, on the shooting modes list, please make sure you click on **PHOTO** and after that, please click ⭕ to take a photo.

Scene optimiser

Also, when the camera is able to recognise the subject, note that the scene optimiser button will change and the optimized colour as well as the effect will be applied.

In case the feature is not activated, please make sure you click ⚙ on the preview screen and please don't forget to also click on the **Scene optimiser** so as to switch to activate it.

Shot suggestions

On the other hand, the camera proposes the ideal composition for the photo by recognizing the position as well as the angle of your subject.

Also, in the preview screen, please don't forget to click on ⚙ and after that, please make sure you click on the **Shot suggestions** switch so as to activate it.

a. While in the shooting modes list, please make sure you click on **PHOTO**.

b. Please note that you are going to receive a guide on the preview screen.

c. Please make sure you point the guide at the subject.

d. Also when the camera recognises the composition, the recommended composition will definitely appears on the preview screen.

e. Please make sure you move the device so that the guide can matches the recommended composition.

f. Also when the ideal composition is finally achieved, the guide will automatically turn to yellow.

g. Please don't forget to click on ◯ to take a photo.

Taking selfies

Please note that you can as well take self-portraits with the front camera through the following method:

1. Meanwhile, on the preview screen, please swipe upwards or downwards or you can as well click

on ⊙ to switch to the front camera for self-portraits.

2. Please make sure you are facing the front of the camera lens.

3. Also, to take self-portraits with a wide-angle shot of the landscape or people, please click on 👥.

4. Please don't forget to click ◯ to take a photo.

Applying filter and beauty effects

Also, it is very possible for you to choose a filter effect as well as a modify facial features, such as your skin tone or face shape, before taking a photo.

1. Meanwhile, on the preview screen, please make sure you click on ✳.

2. Please don't forget to choose the effects along with taking a photo.

In case you want to make use of my filters feature, please note that it is very possible for you to create your own filter using an image with a colour tone that you prefer from **Gallery**.

Locking the focus (AF) and exposure (AE)

Also, it is very possible for you to lock the focus or exposure on a chosen area so as to prevent the camera from automatically adjusting itself based on changes to the subjects or light sources.

Please don't forget to click and hold the area to focus, please note that when the AF/AE frame appear on the area ,the focus as well as exposure setting will be locked. And as such, the setting will still be locked even after you take a photo.

Video mode

Furthermore, the camera adjusts the shooting options automatically depending on the surroundings that is available to record videos easily.

1. Also, on the shooting modes list, please click on **VIDEO** and after that; please make sure you click ⬤ to record a video.

   - Also, to switch between the front as well as rear cameras while recording, please make sure you swipe upwards or downwards on the preview screen or you can as well click on ⊙.

   - Again, to capture an image from the video while recording, please click on ▣.

   - Also, to change the focus while recording a video, please make sure you click on where you want to focus. In case you want to make use of auto focus mode, please click on **AF** to cancel the manually set focus.

■ Please make sure you click to stop recording the video.

Also, the optical zoom may not work in low-light environments.

Stabilising videos (Super steady)

When recording videos, it is very important that you stabilise them through the use of a Super steady feature.

Please don't forget to click on **VIDEO** on the shooting modes list, please make sure you also click ꝥ on the shooting options to activate it, and then record a video.

Single take mode

Please make sure you take various photos as well as videos in just one shot.

Meanwhile, your device can automatically choose the best shot as well as creating photos with filters or videos with certain sections repeated.

1. Also, on the shooting modes list, please click on **SINGLE TAKE**.

2. Please don't forget to click ◯ and move the camera to capture the scene you want.

3. Please when you are done, don't forget to click on the preview thumbnail.

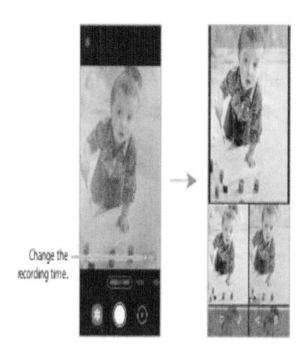

Change the recording time.

Gallery

Introduction

Please note that it is very possible for you to view images as well as videos that are stored in your device. Also, you can manage images as well as videos by album or create stories.

Please don't forget to launch the **Gallery** app.

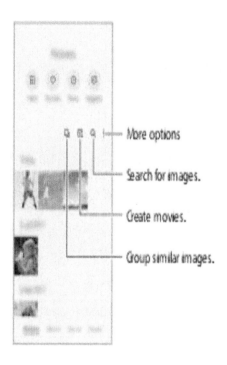

Grouping similar images

Please make sure you launch the **Gallery** app and after that, please click on ⧉ to group similar images as well as displaying only the best shots as a preview of the images. Also, when you click on the image preview, note that it is very possible for you to view all the images in the group.

Creating movies

More so, you can create a movie by choosing a images or videos. Please make sure you launch the **Gallery** app, please don't forget to click on ⌖. Also make sure you tick the files you want to use, and after that, please click on **Create movie**. Please note that when you choose a **Highlight reel**, the device will automatically combine the highlights from the images or videos as well as creating a movie.

Viewing images

Please make sure you launch the **Gallery** app as well as chose an image. In case you want to view other files, please swipe to the left or right on the screen.

Modify the image.

Add the image to favourites.

More options

Bixby Vision

Share the image with others.

Delete the image.

Chapter 4: Samsung Apps

AR Zone

Introduction

Basically, the AR Zone provides you with the AR related features. Please make sure you choose a feature as well as capture fun photos or videos.

Launching AR Zone

The following methods can be use to launch AR Zone:

- Make sure you launch the **AR Zone** app.
- Please don't forget to launch the **Camera** app and click on **MORE** → **AR ZONE**.

AR Emoji Camera

Creating AR Emoji

Meanwhile, you can actually make an emoji that looks like you. Note that emoji stickers with diverse expressions will be generated automatically.

1. Please launch the **AR Zone** app and don't forget to click on **AR Emoji Camera** → ⊕.

2. Please make sure you align your face on the screen, after that please click on ◯ to take a photo, thereafter follow the on-screen instructions to create an emoji.

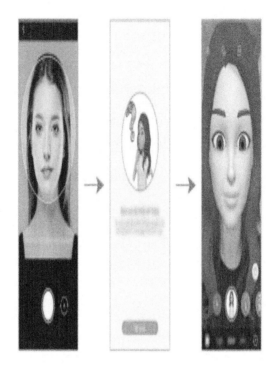

Deleting an emoji

Also, please don't forget to launch the **AR Zone** app and click on **AR Emoji Camera**. Please click on ⚙ → **Manage emojis**, also don't forget to tick the emoji that you want to delete, and thereafter click on **Delete**. Within a short time, the emoji as well as its emoji stickers will be deleted.

Capturing photos or videos with emojis

Also, create fun photos or videos with emojis using various shooting modes.

1. Please make sure you launch the **AR Zone** app and click on **AR Emoji Camera**.

2. Please don't forget to choose the emoji as well as the mode that you want to use. Note that the available modes may vary depending on the emoji that you have chosen.

- **SCENE**: Meanwhile, it is possible for the emoji to mimics your expressions.

- **MASK**: Also, the emoji's face may appear over your face and as such, it may look like you are wearing a mask.
- **MIRROR**: Also, the emoji can mimic your body movements.
- **PLAY**: Again, the emoji can move on a real background.

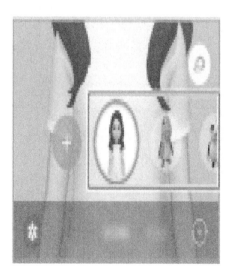

3. Please make sure you click on the emoji icon to take a photo, or you

can as well click and hold the icon to record a video.

Please note that you can view and share the photos as well as videos that you have captured in **Gallery**.

AR Emoji Stickers

Also, it is possible for you to create your own stickers with your emoji's expressions as well as actions. Meanwhile, it is very possible for you to use your emoji stickers when sending messages or on a social network.

Creating your own stickers

1. Please launch the **AR Zone** app and don't forget to click on **AR Emoji Stickers**.
2. Please make sure you click on **Make custom stickers** at the bottom of the screen.

3. Again, you can edit stickers the way you want and please make sure you click on **Save**.

Also, you can actually view the stickers that you have created simply by clicking on **Custom**.

Deleting emoji stickers

Please make sure you launch the **AR Zone** app and also don't forget to click on **AR Emoji Stickers** → **⋮** → **Delete stickers**. Please make sure you choose the emoji stickers to delete plus click on **Delete**.

Using your emoji stickers in chats

Furthermore, please note that you can actually make use of your emoji stickers during a conversation via messages or on a social network. The following actions are instances that you can make use of your emoji stickers in the **Messages** app.

1. Also, when composing a message in the **Messages** app, please click 😊 on the Samsung keyboard.

2. Please don't forget to click on the emoji icon.

3. Please don't forget to also choose one of your emoji stickers.

Please note that the emoji sticker will be inserted.

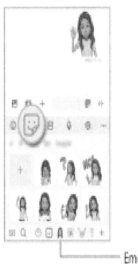

Emoji icon

Bixby

Introduction

Basically, the bixby is a user interface that helps you to make use of your device more conveniently.

Meanwhile, you can actually talk to Bixby or type text. Please note that the bixby can actually launch a function that you have requested or show the information you want.

Again, bixby is only available in some languages, and as such, it may not be available depending on the region.

Starting Bixby

Please make sure you press and hold the Side key to launch Bixby .After that, the bixby intro page will appear. Once you choose the language to use with Bixby, please make sure you sign into your Samsung account, as well as completing the setup simply by following the on-screen instructions. After that, the Bixby screen will appear.

Using Bixby

Also, while pressing as well as holding the Side key, please make sure you say what you want to Bixby, and after that, please release your finger from the key when you are done speaking. Alternatively, please say "Hi, Bixby" and say what you want.

For instance, while pressing as well as holding the Side key, please make sure you say "How's the weather today?" Once you are done with that, the weather information will be displayed on the screen.

Also, if you are interested about tomorrow's weather, just press and hold the Side key, and say "Tomorrow"

But if Bixby asks you a question during a conversation, while pressing and holding the Side key, answer Bixby. Or, you can as well click  and answer Bixby.

In case you are using a headphones or Bluetooth audio devices, you can actually start a conversation by saying "Hi, Bixby", also, you can continue the conversation without clicking the icon. Please make sure you launch the **Bixby** app andalso don't forget to click on →

→ **Automatic listening** → **Hands-free only**.

Waking up Bixby using your voice

Please note that you can actually start a conversation with Bixby simply by saying "Hi, Bixby". Please don't forget to register your voice so that the Bixby can respond to your voice when you say "Hi, Bixby".

1. Please make sure you launch the **Bixby** app and tap → → **Voice wake-up**.

2. Please click on the **Wake with "Hi, Bixby"** switch to activate it.

3. Please make sure you follow the on-screen instructions so as to complete the setup.

At this time you can say "Hi, Bixby" and start a conversation.

Communicating by typing text

Also, you can as well communicate with Bixby via text when your voice isn't recognized or when you are in a noisy environment.

Please make sure you launch the **Bixby** app, don't forget to tap, and afterward type what you want.

Also, during the communication, the Bixby can answer call through text rather than through the voice feedback.

Bixby Vision

Introduction

Basically, the bixby Vision is also a service that offers different features based on image recognition. Also, you can make use of the Bixby Vision to quickly search for information through recognising objects. Please note that you can actually make use of a variety of useful Bixby Vision features.

Launching Bixby Vision

Launch Bixby Vision using one of these methods.

- Also, in the **Camera** app, please click on **MORE** on the shooting modes list and tap **BIXBY VISION**.

- In the **Gallery** app, please click on an image and tap **◉**.

- In the **Internet** app, please don't forget to click and hold an image and tap **Bixby Vision**.

- If you added the Bixby Vision app icon to the Apps screen, please don't forget to launch the **Bixby Vision** app.

1. Please make sure you launch Bixby Vision.

2. Please don't forget to choose a feature that you want to use.

- Meanwhile, you can actually recognise text from documents or images as well as translating it.

- Also, you can as well search for images that is comparable to the recognised object

- Online and related information.

- Please don't forget to search for information based on the products please make sure you also try to recognise QR codes as well as the view information.

Although, the available features as well as the search results may also differ depending on the region or service provider.

Bixby Routines

Introduction

Meanwhile, you can also include your repeated usage patterns as routines as well as making good use of your device conveniently.

For instance, a 'before bed' routine performs actions, such as turning on silent mode as well as dark mode, to not be harsh on your eyes and ears when you use the device before going to bed.

Adding routines

1. Please make sure you launch the **Settings** app and don't forget to click **Advanced features** → **Bixby Routines**.

2. Again, on the **Discover** list, please don't forget to choose a routine that you want or you can as well click on ✚ to add your own routines.

o Meanwhile, you can set the conditions as well as actions of routines on the **Discover** list.

o In case you want to set the routine's running condition to manual, please click on **Start button tapped**. Please note that this option will appear only when there are no running conditions set. Also when a pop-up window appears, please make sure you click on **Add**. Apart from that, you can also add the routine to the Home screen as a widget and access it quickly.

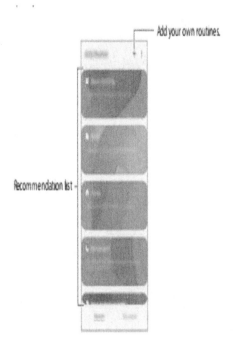

Add your own routines.

Recommendation list

Samsung Health

Introduction

Basically, the samsung Health assist you in managing your wellness as well as fitness. Based on the fact it helps to set fitness goals, check your progress, as well as keeping track of your overall wellness as well as fitness. Also, it is very possible for you can also compare your step count records with other Samsung Health users and view health tips.

Using Samsung Health

Please don't forget to launch the **Samsung Health** app. Also, when running this app for the first time or restart it after performing a data reset, please make sure you follow the on-screen instructions to complete the setup.

Also, to edit items on the Samsung Health home screen, please make sure you click on **Manage items** at the bottom of the trackers list.

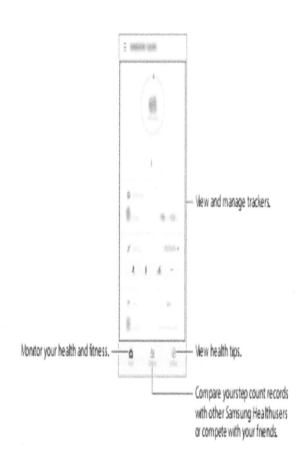

View and manage trackers.

Monitor your health and fitness. — View health tips.

Compare your step count records with other Samsung Health users or compete with your friends.

Galaxy Wearable

Please note that the galaxy Wearable is an app that permits you to manage your wearable devices. Again, when you connect your device to the wearable device, you can actually customise the wearable device's settings as well as apps.

Please don't forget to launch the **Galaxy Wearable** app.

Please click on Get started to connect your device to the wearable device. Please don't forget to follow the on-screen instructions so as to complete the setup. For more information on how to connect and use the wearable device with your device please refer to the wearable device's user manual .

Calendar

Meanwhile, you can manage your schedule by entering upcoming events in your planner.

Creating events

1. Please launch the **Calendar** app and don't forget to click on or double-tap a date.

2. In case the date already has saved events or tasks in it, please click on the date and click .

3. Please don't forget to enter event details and tap **Save**.

Syncing events with your accounts

1. Please make sure you launch the **Settings** app, and click on **Accounts and backup** → **Accounts**, after that please choose the account to sync with.

2. Please don't forget to click on **Sync account** and tap the **Calendar** switch to activate it.

Please for the Samsung account, don't forget to click on $\vdots$ → **Sync settings** and after that, please click the **Calendar** switch to activate it.

Meanwhile, to include accounts to sync with, please launch the Calendar app andplease don't forget to click on →

→ Add new account. After that, please choose an account to sync with and sign in. Please note that a a blue circle is displayed next to the account name when an account is added.

Reminder

You can also register to-do items as reminders and receive notifications according to the condition you set.

- Please make sure you connect to a Wi-Fi or mobile network to receive more accurate notifications.

- For you to utilize the location reminders, the GPS feature must be activated. Although, the location reminders may not be available depending on the model.

Starting Reminder

Please launch the **Calendar** app and sont forget to click on $\equiv$ → **Reminder**. Please note that the Reminder screen will appear and as such, the Reminder app icon (🔔) will be added to the Apps screen.

Creating reminders

1. Please don't forget to launch the **Reminder** app.

2. Please click on **Write a reminder** or ✚, enter the details, and after that click on **Save**.

Completing reminders

Again, on the reminders list, please choose a reminder and click on **Complete**.

Restoring reminders

Please don't forget to restore reminders that have been completed.

1. Meanwhile, on the reminders list, please click on ⋮ → **Completed** → **Edit**.
2. Please don't forget to tick items to restore and tap **Restore**.

Please note that reminders will be added to the reminders list and you will be reminded again.

Deleting reminders

In case you want to delete a reminder, please make sure you choose a reminder and click on **Delete**. Also, to delete multiple reminders, please click and hold a reminder, also tick reminders to delete, and after that please click on **Delete**.

Voice Recorder

- Please don't forget to record or play voice recordings.
- Please make sure you launch the **Voice Recorder** app.
- Please click on start recording after that please speak into the microphone.
- Please don't forget to click on pause recording.
- Also, while making a voice recording, please click on **BOOKMARK** to insert a bookmark.
- Please make sure you click ▊ to finish recording.

- Please don't forget to also enter a file name and tap **Save**.

Changing the recording mode

Please don't forget to choose a mode from the top of the voice recorder screen.

- **Standard**: This is basically the normal recording mode.

- **Interview**: Please note that the device records sound from the top as well as the bottom of the device at a high volume at the same time as reducing the volume of sound from the sides.

- **Speech-to-text**: Please note that the device can record your voice and can simultaneously converts it to on-screen text. Again, for best results, please ensure to keep the device close to your mouth and speak loudly and clearly in a quiet place.

In case the voice memo system language does not match the language you are speaking, note that the device will not recognise your voice. Before using this feature, please make sure you click on the current language to set the voice memo system language.

Playing selected voice recordings

Please note that when you review interview recordings, it is very possible for you to mute or unmute certain sound sources in the recording.

1. Please don't forget to click on **List** to choose a voice recording made in interview mode.

2. In case you want to mute certain sound sources, please click on ON

for the corresponding direction that sound is to be muted.

Please note that the icon will change to OFF and the sound will be muted.

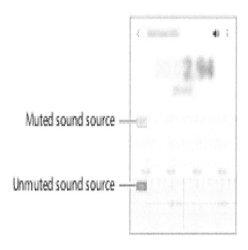

My Files

Please make sure you access and manage various files stored in the device.

Please don't forget to launch the **My Files** app.

In case you want to check for unnecessary data and free up the device's storage, please click on **Analyse storage**. To search for files or folders, please make sure you also click on Q.

Clock

It is also possible for you to set alarms, check the current time in many cities around the world, time an event, or set a specific duration. Please make sure you launch the **Clock** app.

Calculator

Please note that the calculator can perform simple or complex calculations.

Please make sure you launch the **Calculator** app.

⏱ : You can also view the calculation history. You can also clear the history, please click on**Clear history**. Tin case you want to close the calculation history panel, please click on ⊞.

▭ : Please make sure you make use of the unit conversion tool. It is also possible for you to convert various values, such as area, length, or temperature, into other units.

⊡ : It is also possible for you to display the scientific calculator.

Game Launcher

Please note that the game Launcher gathers the games downloaded from **Play Store** as well as from the **Galaxy Store** into another place for easy access. And it is also possible for you to set the device to game mode for the purpose of easy playing of games.

Please make sure you launch the **Game Launcher** app and please don't forget to choose the game you want.

- In case the **Game Launcher** does not appear, please launch the **Settings** app and click on **Advanced features**, and after that please click on the **Game Launcher** switch to activate it.

- Please note that the games downloaded from **Play Store** and **Galaxy Store** will be displayed automatically on the Game Launcher screen. But if you can't see your games, make sure you drag the Library panel upwards and click on ⁝ → **Add apps**.

Removing a game from Game Launcher

Please don't forget to drag the Library panel upwards, please click and hold a game, and after that please click on **Remove from Game Launcher**.

Changing the performance mode

The following are ways you can change the game performance mode.

Please launch the **Game Launcher** app, and click on ▓ → **Game performance** → **Game performance**, after that, please choose the mode you want.

- **Focus on performance**: This is based on giving you the best possible performance while playing games.
- **Balanced**: This on the other hand, helps to balance the performance as well as the battery usage time.
- **Focus on power saving**: This focus on saving battery power while playing games.

Although, battery power efficiency may differ by game.

Game Booster

Also, the game Booster allows you to play games in a better environment and as such, you can make use of the Game Booster while playing games.

Also, to open the Game Booster panel while playing games, please click ⚬ on the navigation bar. In case the the navigation bar is hidden, please make sure you drag upwards from the bottom of the screen to show it. But if you have set the navigation bar to use **Swipe gestures**, please open the notification panel and don't forget to click **Tap to open Game Booster.**

⚙: Also, make sure you configure settings for Game Booster.

Monitoring temperature / Monitoring memory: Furthermore, you can set the device to automatically adjust settings so as to prevent the device from overheating as well as to prevent apps running in the background to better manage memory.

Block during game: Again, you can also lock some features during games.

Navigation button lock: You can also hide the buttons on the navigation bar. In case you want to display the buttons, please click on the navigation bar.

Screen touch lock: Please make sure you lock the touchscreen while the game is being played. Also, to unlock the touchscreen, please drag the lock icon in any direction.

Screenshot: Please don't forget to capture screenshots.

Again, it is possible for you to open the Game Booster panel from the navigation bar while the navigation bar is set to **Swipe gestures**. Directly from the the Game Booster panel, please click on **Block during game** and make sure you click on the **Navigation gestures** switch to activate it.

Please note that available options may differ depending on the game.

Launching apps in pop-up windows while playing games

It is important to note that you can launch apps in pop-up windows while playing a game.

Please don't forget to click and choose an app from the apps list.

Samsung Kids

More so, it is possible for you to restrict children's access to certain apps, set their usage times, as well as configure settings that will provide a fun as well as a safe environment for children when they use the device.

Please make sure you open the notification panel, and please don't forget to swipe downwards, and after that please click on (**Samsung Kids**) to activate it. Immediately, the Samsung Kids screen will appear. Also, when starting Samsung Kids for the first time or after performing a data reset, please make sure you follow the on-screen instructions to complete the setup.

Directly from the Samsung Kids screen, please choose the app you want to use.

Please note that your preset screen lock method or your created PIN will be used when activating the **Parental control** feature or closing Samsung Kids.

Using parental control features

Please note that it is very possible for you to configure the settings for Samsung Kids as well as view the usage history.

Directly from the Samsung Kids screen, please click on ⋮ → **Parental control** and enter your unlock code.

Closing Samsung Kids

In case you want to close Samsung Kids, please click on the Back button or you can as well click on ⋮ → **Close Samsung Kids**, and after that , please enter your unlock code.

SmartThings

Also, you can control as well as manage smart appliances and Internet of Things (IoT) products with your smartphone.

In case you want to view more information, please launch the **SmartThings** app and click on ☰ → **How to use**.

1. Please make sure you launch the **SmartThings** app.

2. Please don't forget to click on ✚ → **Device**.

3. Please make sure you choose a device and connect to it by following the on-screen instructions.

- Meanwhile, the connection methods may vary depending on the type of connected devices or the shared content.

- And as such, the devices can connect to may also differ depending on the region; again, the Available features may differ depending on the connected device.

- Please note that Samsung warranty do not cover connected devices' own errors or defects. Also, when errors or defects occur on the connected devices, please make sure you contact the device's manufacturer.

Sharing content

Meanwhile, you can also share content by using various sharing options. Below are some of the ways you can share images.

1. Please make sure you launch the **Gallery** app as well as to choose an image.

2. Pease don't forget to click as well as to choose a sharing method you want.

3. Please note that you may incur additional charges when sharing files via the mobile network.

Quick Share

Sharing content with nearby devices

However, you can share content with nearby devices via Wi-Fi Direct or Bluetooth, or with SmartThings supported devices through the following method:

1. Please make sure you launch the **Gallery** app as well as choose an image.

2. Directly from the other device, please open the notification panel, and don't forget to swipe downwards. After that please click ⊚ (**Quick Share**) to activate it.

3. Please don't forget to click ⤳ and decide on the device to transfer the image to.

4. Please make sure you accept the file transfer request on the other device.

Setting who can find your device

The following are the ways you can set who is allowed to find and send content to your device:

1. Please open the notification panel and swipe downwards. After that ,please click (**Quick Share**) to activate it.

2. Please don't forget to click and hold (**Quick Share**).

3. Immediately after that, the Quick Share settings screen will appear.

4. Please make sure you choose an option.

   - **Contacts only**: Please don't forget to permit only your contacts to share with your device.

   - **Everyone**: Please make sure you permit any nearby devices to share with your device.

Link Sharing

The following are the ways you can share large files as well as upload files to the Samsung storage server and share them with others via a Web link:

1. Please, make sure you lunch the **Gallery** app and choose an image.
2. Please don't forget to click ⚷ → Link Sharing.
3. Immediately, the link for the image will be created.
4. Please don't forget to choose a sharing option.

Shared album

The following are the ways that you can create a shared album to share photos or videos with others, and download your files whenever you want:

1. Please launch the **Gallery** app and select an image.

2. Please make sure you click on ⦵ → Shared album.

3. Please choose an album to share.

- In case there is no album to share, please click on **Create shared album** and follow the on-screen instructions to create an album.

- Again, when you play a high-resolution video from a shared album, the video connection may drop depending on the network speed.

- Also, the content that is larger than 1 GB cannot be shared to a shared album.

Music Share

Introduction

Again, the Music Share feature permits you to share your Bluetooth speaker that is already connected to your device with another person. You can also listen to the same music on your

Galaxy Buds as well as another person's Galaxy Buds.

Please note that this feature is available only on devices that support the Music Share feature.

Sharing a Bluetooth speaker

Basically, it is possible for you can listen to music on your smartphone and your friend's smartphone via your Bluetooth speaker.

1. Please make sure that your smartphone and your Bluetooth speaker are connected. You can also refer to Pairing with other Bluetooth devices for how to connect.

2. Directly from your smartphone, open the notification panel, please swipe downwards, and after that check if 🎵 (**Music Share**) is activated.

3. Also, you can make use of additional features, such as setting who to share your

device with, simply by tapping and holding

(**Music Share**).

4. Directly from your friend's smartphone, please choose your speaker from the list of the Bluetooth devices.

5. Again, directly from your smartphone, please accept the connection request.

6. Please note that your speaker will be shared. Also, when you play music via your friend's smartphone, the music playing via your smartphone will be paused.

Listening to music together with Galaxy Buds

- Again, it is possible for you can listen to music on your smartphone together through your Buds and your friend's Buds.

- Please note that this feature is supported only on Galaxy Buds, Galaxy Buds Plus, as well as Galaxy Buds Live.

1. Please make sure that each smartphone and pair of Buds is connected.

2. Please don't forget to refer to Pairing with other Bluetooth devices for how to connect.

3. Directly from your friend's smartphone, please open the notification panel, swipe downwards, and after that please click (**Music Share**) to activate it.

4. Also, you can make use of additional features, such as setting who to share your device with, simply by tapping and holding (**Music Share**).

5. Directly from your smartphone, please open the notification panel and click on **Media**.

6. Please don't forget to click ♫ and choose your friend's Buds from the detected devices list.

7. Also, on your friend's smartphone, please don't forget to accept the connection request.

8. Directly from your smartphone, please make sure you tick your Buds and your friend's Buds on the audio output list.

9. Also, when you play music via your smartphone, note that you can listen to it together via both Buds.

Chapter 5: Mobile Continuity

Samsung DeX

Introduction

Meanwhile, the samsung DeX is a service that permits you to make use of your smartphone like a computer simply by connecting the smartphone to an external display, such as a TV or monitor, or to a computer. Also, you can accomplish tasks you want done on your smartphone quickly as well as easily on a large screen using a keyboard as well as mouse. Also, while using Samsung DeX, it is also possible for you to simultaneously make use of your smartphone.

Wired connections to external displays

Also, it is very possible for you to connect your smartphone to an external display when using an HDMI adaptor (USB Type-C to HDMI).

- Please make sure you make use of only official Samsung DeX supported accessories that are provided by Samsung. Please note that performance problems as well as malfunctions that are caused by using accessories that are not officially supported are not covered by the warranty.

1. Please don't forget to also connect an HDMI adaptor to your smartphone.

2. Also, please connect an HDMI cable to the HDMI adaptor and to a TV or monitor's HDMI port.

3. Directly from your smartphone's screen, please click on **Start**.

Again, when you don't change your smartphone's screen, the Samsung DeX screen will appear on the connected TV or monitor.

Wired connections to computers

1. Also, you can make use of the Samsung DeX simply by connecting your smartphone to a computer using a USB cable.

2. Please don't forget to also connect your smartphone to a computer using a USB cable.

3. Directly from your Smartphone's screen, please click on **Start now**.

As soon as they are connected, the Samsung DeX screen will appear on the computer.

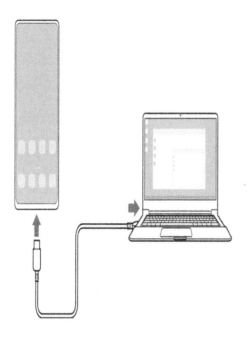

Connecting to a TV wirelessly

Also, you can make use of the Samsung DeX by connecting your smartphone to a TV wirelessly.

1. Directly from your smartphone, please open the notification panel and don't forget to swipe downwards. After that, please click on 🖼 (**DeX**).

2. Please don't forget to choose a TV from the detected devices list and please make sure you click on **Start now**.

3. Please note that some TVs will only be detected when the screen mirroring mode is turned on.

4. In case the connection request window appears on the TV, please make sure you accept the request.

5. Please don't forget to also follow the on-screen instructions to complete the connection.

6. Meanwhile, when they are connected, note that the Samsung DeX screen will come into view on the TV.

- You are therefore advised to make use of a Samsung Smart TV manufactured after 2019.

- Please make sure that screen mirroring is supported on the TV you want to connect to.

Controlling the Samsung DeX screen

Controlling on an external display

Controlling with an external keyboard and mouse Furthermore, note that it is possible for you to make use of a wireless keyboard/mouse.

- Apart from that, you can also set the mouse pointer to flow from the external display to the smartphone's screen. Please don't forget to lunch **settings** app, thereafter choose the Samsung **DeX** → **Mouse/trackpad**, and then decide on the **Flow pointer in the phone screen** switch to activate it.

- Please note that you can also make use of the external keyboard on the smartphone's screen.

Using your smartphone as a touchpad

The following are the ways you can use your smartphone as a touchpad:

Please directly from your smartphone, please drag downwards from the top of the screen to open the notification panel and after that, please click on Use **your phone as a touchpad**.

- Also, when you open the notification panel, you can actually view the gestures that you want to use with the touchpad.

- In case your smartphone's case has a front cover, please open the front cover to use your smartphone as a touchpad. But if the front cover is closed, please note that the touchpad may not work properly.

- Also, when your smartphone's screen turns off, please don't forget to press the Side key or double-tap the screen to turn on the screen.

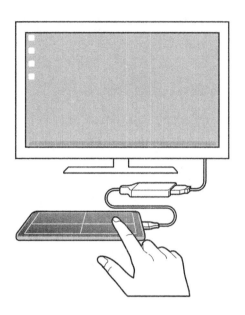

Chapter 6: Settings

Customize device settings.

Please launch the **Settings** app.

In case you want to search for settings by entering keywords, please click on Q.

Connections

Options

- Please make sure you change the settings for various connections, such as the Wi-Fi feature and Bluetooth.

- Again, directly from the Settings screen, please click on **Connections**.

- **Wi-Fi**: Meanwhile, you can activate the Wi-Fi feature to connect to a Wi-Fi network and access the Internet or other network devices.

- **Bluetooth**: Please note that you can make use Bluetooth to exchange data or media files with other Bluetooth-enabled devices.

- **NFC and payment**: Also, you can actually set the device to permit you to read near field communication (NFC) tags that contain information about products. Apart from that, you can also make use of this feature to make payments as well as buy tickets for transportation or events after downloading the required apps.

- **Flight mode**: Again, to set the device to disable all wireless functions on your device, please note that you can only make use of non-network services.

  ⓘ Please make sure you follow the regulations provided by the airline as well as the instructions of aircraft personnel. Again, in cases where it is allowed to use the device, please make sure you always use it in flight mode.

- **Mobile networks**: Please don't forget to configure your mobile network settings.

- **Data usage**: Please make sure you keep track of your data usage amount as well as customise the settings for the limitation. Please don't forget to also set the device to automatically disable the mobile data connection when the amount of mobile data you have used reaches your specified limit.

- Meanwhile, you can activate the data saver feature to prevent some apps running in the background from sending or receiving data.

- Apart from that, you can also choose apps to always use the mobile data even when your device is connected to a Wi-Fi network.

- **SIM card manager** (dual SIM models): please don't forget to activate your SIM or USIM cards and customise the SIM card settings.

- **Mobile Hotspot and Tethering**: Please make sure you use the device as a mobile hotspot to share the device's mobile data connection with other devices.

- Please note that you may incur additional charges when using this feature.

- **More connection settings**: Please make sure you customise settings to control other features.

Wi-Fi

Please don't forget to activate the Wi-Fi feature to connect to a Wi-Fi network as well as to access the Internet or other network devices.

Connecting to a Wi-Fi network

1. Directly from the Settings screen, please click on **Connections** → **Wi-Fi** and please don't forget to switch to activate it.

2. Please make sure you choose a network from the Wi-Fi networks list.

Please note that the networks with a lock icon require a password.

- As soon as the device connects to a Wi-Fi network, please note that the device will reconnect to that network each time it is available without requiring a password. To prevent the device from connecting to the network automatically, please make sure you click on ⚙ next to the network and after that; please click on the **Auto reconnect** switch to deactivate it.

- Also, if you cannot connect to a Wi-Fi network properly, please make sure you restart your device's Wi-Fi feature or the wireless router.

Viewing the Wi-Fi network quality information

Meanwhile, the following are the ways you can view the Wi-Fi network quality information, such as the speed and stability.

- Directly from the Settings screen, please click on **C uponnections** → **Wi-Fi** and please make sure you click the switch to activate it. Also, the network quality information will appear under the Wi-Fi networks. In case it t does not appear, please click on ⋮ → **Advanced** and tap the **Show network quality info** switch to activate it.

- Please note that the quality information may not appear depending on the Wi-Fi network.

Sharing Wi-Fi network passwords

In case you make a request to a person who is connected to a secured Wi-Fi network to share its password, please note that you can connect to the network without entering the password. Please be informed that this feature is available between the devices which have contacts each other and the screen of the other device must be turned on.

1. Directly from the Settings screen, please click on **Connections** → **Wi-Fi** and please don't forget to switch to activate it.

2. Please make sure you choose a network from the Wi-Fi networks list.

3. Please don't forget to click on Request password.

4. Please accept the share request on the other device.

Also, the Wi-Fi password is entered on your device and it is connected to the network.

Wi-Fi Direct

Please note that the Wi-Fi Direct connects devices directly via a Wi-Fi network without requiring an access point.

1. Directly from the Settings screen, tap **Connections** → **Wi-Fi** and click on the switch to activate it.

2. Please don't forget to click on ⋮ → Wi-Fi Direct.

   - Please note that the detected devices are listed.

   - In case the device you want to connect to is not in the list, please make sure you request that the device turns on its Wi-Fi Direct feature.

3. Please don't forget to choose a device to connect to.

   Please note that the devices will be connected when the other device accepts the Wi-Fi Direct connection request.

   Also, to end the device connection, please make sure you choose the device to disconnect from the list.

Sending and receiving data

Please note that you can also share data, such as contacts or media files, with other devices. The following are ways of sending an image to another device:

1. Please launch the **Gallery** app and choose an image.

2. Please make sure you click ⦿ → **Wi-Fi Direct** and select a device to transfer the image to.

3. Please don't forget to acccept the Wi-Fi Direct connection request on the other device.

Also, if the devices are already connected, the image will be sent to the other device without the connection request procedure.

Bluetooth

Please don't forget to make use Bluetooth to exchange data or media files with other Bluetooth-enabled devices.

- ⓘ Please note that samsung is not responsible for the loss, interception, or misuse of data sent or received via Bluetooth.

- Please make sure you always ensure that you share and receive data with devices that are trusted and properly secured. However, if there are obstacles between the devices, the operating distance may be reduced.

- Please note that some devices, especially those that are not tested or approved by the Bluetooth SIG, may be incompatible with your device.

Pairing with other Bluetooth devices

1. Also, on the Settings screen, please click on **Connections** → **Bluetooth** and tap the switch to activate it. The detected devices will be listed.

2. Please don't forget to choose a device to pair with.

- In case the device you want to pair with is not on the list, please make sure you set the device to enter Bluetooth pairing mode.

- Please note that your device is visible to other devices while the Bluetooth settings screen is open.

3. Please don't forget to accept the Bluetooth connection request on your device to confirm.

Also, the devices will be connected when the other device accepts the Bluetooth connection request.

Again if you want to unpair the devices, please click on ⚙ next to the device name to unpair and click **Unpair**.

Sending and receiving data

Also, many apps support data transfer via Bluetooth. But you can till share data, such as contacts or media files, with other Bluetooth devices.

The following are ways of sending image to another device:

1. Please launch the **Gallery** app and choose an image.
2. Please don't forget to click ⚬⟨ → **Bluetooth** and select a device to transfer the image to.

In case the device you want to pair with is not in the list, please make request that the device is visibility option.

3. Please don't forget to accept the Bluetooth connection request on the other device.

Mobile data only apps

Please make sure you choose a apps to always use the mobile data even when your device is connected to a Wi-Fi network.

For instance, you can set the device to use only mobile data for apps that you want to keep secure or streaming apps that can be disconnected. Also, even if you do not deactivate the Wi-Fi feature, note that the apps will launch using the mobile data.

Directly from the Settings screen, please click on **Connections → Data usage → Mobile data only apps**, please make sure you also click on the switch to activate it, as well as clicking on the switches next to the apps .

Mobile Hotspot

Meanwhile, you can make use of your device as a mobile hotspot to share your device's mobile data connection with other devices.

1. Also, directly from the Settings screen, please click on Connections → Mobile Hotspot and Tethering → Mobile Hotspot.
2. Please don't forget to click on the switch to activate it.

- Please note that the 📶 icon appears on the status bar.

- Also it is very possible for you to change the level of security and the password simply by tapping ⋮ → **Configure Mobile Hotspot**.

3. Again, directly from the other device's screen, search for and please don't forget to please don't forget to choose your device from the Wi-Fi networks list. On the other hand, please click ▦ on the mobile hotspot screen and scan the QR code with the other device.

  - In case the mobile hotspot is not found on your device, please set **Band** to 2.4 **GHz**, and don't forget to click on ⋮ → **Configure Mobile Hotspot**, and then deselect **Hide my device**.

  - In case you activate the **Auto Hotspot** feature, please note that you can share your device's mobile data connection with other devices signed in to your Samsung account.

More connection settings

Meanwhile, you can customise settings to control other connection features through:

Directly from the Settings screen, please click on Connections → More connection settings.

- **Nearby device scanning**: Please make sure you set the device to scan for nearby devices to connect to.

- **Printing**: Please don't forget to configure settings for printer plug-ins installed on the device.

- **VPN**: Please make sure you set up virtual private networks (VPNs) on your device to connect to a school or company's private network.

- **Private DNS**: Please make sure you set the device to use the security enhanced private DNS.

- **Ethernet**: Also, when you connect an Ethernet adaptor, note that it is very possible for you to make use of a wired network and configure network settings.

Printing

Please don't forget to configure settings for printer plug-ins installed on the device. However, you can also connect the device to a printer via Wi-Fi or Wi-Fi Direct, as well as print images or documents.

Please note that some of the printers may not be compatible with the device.

Adding printer plug-ins

Please don't forget to add printer plug-ins for printers you want to connect the device to.

1. Directly from the Settings screen, please click on the Connections → More connection settings → Printing → Download plugin.

2. Please don't forget to choose a printer plug-in and install it.

3. Please make sure you choose the installed printer plug-in.

4. Also, the device will be automatically search for printers that are connected to the same Wi-Fi network as your device.

5. Please make sure you choose a printer to add.

Please don't forget to include printers manually as well as click **⋮** → **Add printer**.

Printing content

Also, while viewing content, such as images or documents please make sure you access the options list, please don't forget to click **Print** → **▼** → **All printers...**, and after that choose a printer.

Sounds and vibration

Options

Please note that you can change settings for various sounds on the device.

Directly from the Settings screen, please click on **Sounds and vibration**.

- **Sound mode**: Please set the device to use sound mode, vibration mode, or silent mode.

- **Vibrate while ringing**: please don't forget to also set the device to vibrate and play a ringtone for incoming calls.

- **Temporary mute**: Please make sure you set the device to use silent mode for a certain period.

- **Ringtone**: Please change the call ringtone.

- **Notification sound**: Please don't forget to change the notification sound.

- **System sound**: Please do not forget to change the sound to use for certain actions, such as charging the device.

- **Volume**: Please make sure you adjust the device's volume level.

- **Vibration pattern**: Please don't forget to choose a vibration pattern.

- **Vibration intensity**: Pease make sure you adjust the force of the vibration notification.

- **System sound/vibration control**: Please set the device to sound or vibrate for actions, such as controlling the touchscreen.

- **Sound quality and effects**: Please make sure you set the device's sound quality and effects.

- **Separate app sound**: Please don't forget to set the device to play media sound from a specific app separately on the other audio device.

Sound quality and effects

Please make sure you set the device's sound quality and effects.

Directly from the Settings screen, please click on the Sounds and vibration → Sound quality and effects.

- **Dolby Atmos**: Please make sure you choose a surround sound mode optimised for various types of audio, such as movies, music, and voice. **Dolby Atmos for gaming**: Meanwhile, you can experience the Dolby Atmos sound optimised for games while playing games.
- **Equaliser**: Please make sure you choose an option for a specific music genre and enjoy optimised sound.
- **UHQ upscaler**: Please don't forget to also enhance the sound resolution of music and videos.

- **Adapt sound**: Please make sure you set the best sound for you.

Separate app sound

Please don't forget to set the device to play media sound from a specific app on the connected Bluetooth speaker or headset.

For instance, it is very possible for you to listen to the Navigation app through your device's speaker while listening to playback from the Music app through the vehicle's Bluetooth speaker.

1. Again, directly from the Settings screen, please click on **Sounds and vibration** → **Separate app sound** and please don't forget to also click on the switch to activate it.

2. Please don't forget to also choose an app to play media sounds separately and tap the Back button.

3. Please make sure you choose a device for playing the selected app's media sound.

Notifications

Meanwhile, you can change the notification settings through:

Directly from the Settings screen, please click on **Notifications**.

- **Suggest actions and replies**: Please make sure you set the device to suggest actions and replies for notifications.
- **Swipe left or right for snooze**: Please don't forget to set the device to show the notification snooze icon when you swipe a notification to the left or right on the notification panel.
- **App icon badges**: Please set the device to display a number or dot badge on apps that have notifications.
- **Status bar**: Please do well to set how to display notification icons and whether to show the remaining battery percentage on the status bar.

- **Do not disturb**: Please make sure you set the device to mute all sounds except for allowed exceptions.

- **Recently sent**: Please don't forget to also view the apps that received recent notifications and change the notification settings. Also, to customise notification settings for more apps, please click on **See all** → ▼ → **All** and choose an app from the apps list.

Display

Options

Please don't forget to change the display and the Home screen settings through:

Directly from the Settings screen, tap **Display**.

- **Light / Dark**: Please don't forget to activate or deactivate dark mode.

- **Dark mode settings**: Please make sure you reduce eye strain by applying the dark theme when using the device at night or in a dark place. Also it is very possible for you to set a schedule for applying dark mode.

- Please note that the dark theme may not be applied in some apps.

- **Brightness**: Please don't forget to adjust the brightness of the display.

- **Adaptive brightness**: Please make sure you set the device to keep track of your brightness adjustments and apply them automatically in similar lighting conditions.

- **Motion smoothness**: Please change the refresh rate of the screen.

- **Blue light filter**: Please make sure you reduce eye strain by limiting the amount of blue light emitted by the screen.

- **Screen mode**: Also, don't forget to change the screen mode to adjust the display's colour and contrast.

- **Font size and style**: Please don't forget to change the font size and style.

- **Screen zoom**: Also don't forget to make the items on the screen larger or smaller.

- **Full screen apps**: Please make sure you choose a apps to use with the full screen aspect ratio.

- **Screen timeout**: Please don't forget to set the length of time the device waits before turning off the display's backlight.

- **Home screen**: Please make sure you configure settings for the Home screen, such as the screen layout.

- **Easy mode**: Please don't forget to also switch to easy mode to display larger icons and apply a simpler layout to the Home screen.

- **Edge screen**: Please make sure you also change the settings for the Edge panel and the edge lighting.

- **Navigation bar**: Please don't forget to change the navigation bar settings.

- **The Accidental touch protection**: Please make sure you set the device so as to prevent the screen from detecting touch input when it is in a dark place, such as a pocket or bag.

- **Touch sensitivity**: Please make sure you increase the touch sensitivity of the screen for use with screen protectors.

- **Show charging information**: Please set the device to display the charging information, such as the remaining battery percentage when the screen is off.

- **Screensaver**: Please make sure you set the device to launch a screensaver when the device is charging.

Motion smoothness

Again, please note that the refresh rate is the number of times the screen is refreshed every second. And as such, please make sure that you make use of a high refresh rate to prevent the screen from flickering when switching between screens. Also, the screen will scroll more smoothly. But when you select a standard refresh rate, you can actually make use of the battery longer.

1. Meanwhile, directly from the Settings screen, please click on **Display** → **Motion smoothness**.

2. Please don't forget to choose a refresh rate.

   - **High**: Please make sure you also get smoother animations and scrolling by automatically adjusting your screen refresh rate up to 120 Hz.

   - **Standard**: Please don't forget to also use a standard refresh rate in normal situations to conserve battery power.

How to Change the screen mode or adjusting the display colour

Please make sure you change the screen mode or adjust the display colour to your preference.

Changing the screen mode

Directly from the Settings screen, please make sure you click on **Display** → **Screen mode** and don't forget to also choose a mode you want.

Vivid: Please note that this optimises the colour range, saturation, as well as sharpness of your display. Although, you can also adjust the display colour balance by colour value.

Natural: This is based on adjusting the screen to a natural tone.

Meanwhile, you can adjust the display colour only in **vivid** mode.

Although, the vivid mode may not be compatible with third-party apps.

Please note that it is not possible to change the screen mode while applying the blue light filter.

Optimising the full screen colour balance

Meanwhile, you can optimise the display colour by adjusting the colour tones to your preference. On the Settings screen, please don't forget to click on **Display** → **Screen mode** → **Vivid** as well as adjusting the colour adjustment bar under **White balance**.

Also, when you drag the colour adjustment bar towards **Cool**, please be inform that the blue colour tone will increase. Again, when you drag the bar towards **Warm**, note that the red colour tone will increase.

Lock screen

Options

Please don't forget to change the settings for the locked screen.

Also, on the Settings screen, please click on **Lock screen**.

Screen lock type: Please don't forget to change the screen lock method.

> **Smart Lock**: Please set the device to unlock itself when trusted locations or devices are detected.

> **Secure lock settings**: Please make sure you change screen lock settings for the selected lock method.

> **Always On Display**: Please set the device to display information while the screen is turned off.

> **Wallpaper services**: Please don't forget to set the device to use wallpaper services such as Dynamic Lock screen.

> **Clock style**: Please don't forget to change the type and colour of the clock on the locked screen.

Roaming clock: Please make sure you change the clock to show both the local and home time zones on the locked screen when roaming.

Face Widgets: Please don't forge to also change the settings of the items displayed on the locked screen.

Contact information: Please make sure you set the device to show contact information, such as your email address, on the locked screen.

Notifications: Please don't forget to set how to show notifications on the locked screen.

Shortcuts: Please make sure you choose the apps to display shortcuts to them on the locked screen.

About Lock screen: Please don't forget to also view the Lock screen version and legal information.

Printed in Great Britain
by Amazon